AEI LEGISLATIVE ANALYSES
Balanced analyses of current proposals before the Congress, prepared with the help of specialists in law, economics, and government

Reauthorization of the Clean Water Act

1983
98th Congress
1st Session

AMERICAN ENTERPRISE INSTITUTE
FOR PUBLIC POLICY RESEARCH
Washington and London

ISBN 0-8447-0248-X
Legislative Analysis No. 33, 98th Congress
July 1983

CONTENTS

1

INTRODUCTION

The federal government has spent more to aid in restoring and maintaining the quality of our surface waters than it has in any other U.S. environmental program. The grant program for building and improving publicly owned wastewater treatment plants is the nation's largest public works program. And the regulatory features of the clean water program are the most litigated of any environmental program. Half again as many suits have been filed under the clean water legislation as under the massive Clean Air Act program.

The thrust of the federal effort to assist in cleaning up our waterways was changed radically with the adoption of the 1972 Federal Pollution Control Act. The goals of that legislation were far more ambitious than those of prior law. Enforcement strategy called for nationally uniform technology-based effluent standards rather than standards based on the quality of receiving waters. Point source dischargers were required to obtain discharge permits, and discharges contrary to the terms of these permits were rendered unlawful. Federal involvement in the administration and enforcement of clean water requirements was far more pervasive than before.

The Clean Water Act, apart from the title II federal grant program for municipal wastewater treatment plants reauthorized in 1981, was due to be reauthorized in 1982, but Congress ran out of time for consideration of proposed changes. The administration's 1983 proposals for change are relatively minor. The bills introduced in the Congress do not follow the administration's proposals in several respects. This analysis will examine the principal areas of dispute and the principal features of pending proposals for change.

2

BACKGROUND

EARLY LEGISLATION

The first federal effort to control industrial water pollution began with the 1948 Water Pollution Control Act, directed at reducing pollution in interstate and coastal waters and in the Great Lakes. It permitted federal enforcement actions if the public health or welfare was endangered, although the states could forestall such suits. Federal loans for the construction of municipal wastewater treatment plants were authorized, but funds were not appropriated for loans until 1956. The states' veto power over enforcement actions was eliminated in the 1956 amendments, but enforcement authorities still had the burden of showing that a particular discharger's effluents were a nuisance. The 1961 amendments extended federal enforcement jurisdiction and expanded grant authority.

The Water Quality Act of 1965 provided for the establishment and enforcement of water quality standards for interstate waters. Breach of the standards by a discharger was sufficient to sustain an enforcement suit. Each state could determine maximum permissible pollution discharges consistent with the water quality standards and could license dischargers. This legislation established the Federal Water Pollution Control Administration and increased sewer grant funds. In 1970 a reorganization plan centralized administrative and enforcement authority in the Environmental Protection Agency (EPA). Congress adopted legislation in that year to control oil spill pollution and sewage discharge from vessels. The 1970 amendments imposed a certification requirement (for activities that otherwise required a federal permit) establishing that discharges would not violate applicable water quality standards. EPA's permit program, modeled on the Refuse Act of 1899 by the plan that created EPA, was limited by court decision to discharge of pollutants into navigable waters.

THE 1972 ACT AND ITS AMENDMENTS

The 1972 Water Pollution Act Amendments replaced a system of pollution controls that had been based on water quality standards (analogous to the ambient air standards still employed under the Clean Air Act), with technology-based effluent standards to be applied uniformly throughout the nation for plants within the same subclass of a particular industry. The new standards required point sources discharging effluents into surface waters to obtain a permit that set out the technology-based standards—standards achievable by employing known technology—for that particular source as related to the beneficial use the state set for the body of receiving water (drinking, recreation, and so forth). Discharges in excess of permit limits were made unlawful. Industrial sources were required to meet best practicable control technology (BCT) by July 1, 1977, and the requirements of best available technology (BAT) by July 1, 1983. Publicly owned wastewater treatment plants were to meet effluent limitations based on "secondary treatment" by July 1, 1977.[1]

Within one year EPA was required to promulgate new source performance standards that represented the greatest degree of effluent reduction achievable by applying the best available demonstrated control technology. The legislation set the ambitious goal of fishable and swimmable water by July 1, 1983, and that was only an interim goal. A zero rate of discharge of pollutants was to be achieved by 1985. Effluent standards based on water quality—controls more stringent than technology-based standards—could be imposed if effluents under the latter standards would interfere with attaining the quality of water that would protect public water supplies, agricultural and industrial uses, and a balanced indigenous population of shellfish, fish, and wildlife, and that would allow recreational activities in and on the water.

EPA was to issue regulations setting pretreatment standards for effluents of industrial plants that would be deposited in publicly owned wastewater treatment plants, including pollutants that were not susceptible to treatment in those plants or that would interfere with the operation of those plants. EPA was also to develop a list of toxic pollutants for control. The 1972 program was federally oriented, while previous programs were state oriented. The legislation established heavier fines for violations and streamlined enforcement. The grant program for municipal wastewater treatment plants authorized payments of $18 billion for a three-year period, and the federal share was increased from 55 percent to 75 percent. EPA administered the permit program for all who discharged wastes into surface waters; qualified states could assume this responsibility, but EPA would still have a veto as to specific permits. EPA was to develop guidelines and issue permits for discharges into the waters of the territorial sea, the contiguous zone, and the ocean. Permits were also required if sludge from sewer plants were to be disposed of in a manner that permitted a pollutant to enter the surface waters. The secretary of the army would issue permits for disposal of dredged or fill material at specified disposal sites.

The Clean Water Act of 1977 reauthorized the popular grant program, and the federal share for projects using innovative technology was increased to 85 percent. Areawide treatment plans were to include consideration of nonpoint sources of pollution such as agricultural runoff. If EPA objected to a state's issuance of a permit, the objection would have to include a description of the effluent limitations and conditions that meet EPA's requirements. The secretary of the army would have to publish a notice and provide opportunity for a hearing on an application for a dredged and fill permit. The secretary was also authorized to issue general permits on a state, regional, or national basis for up to five years.

Congress reauthorized the sewer grant program in 1981 and tightened up on the uses that could be made of grant funds. Grants are authorized beginning in 1985 for the correction of combined storm water and sanitary sewer overflows. The federal share for future projects would be cut from 75 percent to 55 percent. Provisions were included in the 1981 legislation to encourage more economical wastewater treatment plants and to provide greater certainty that completed plants will operate properly.

3

THE ADMINISTRATION'S REAUTHORIZATION BILL

The administration's 1983 clean water bill would reauthorize the program through fiscal 1988. The bill, however, would not change the basic regulatory framework, and technology-based standards would still be employed for point source dischargers. Proposed changes are cited in the following discussion.

The pretreatment program for industrial users who discharge wastes through publicly owned wastewater treatment plants is retained. These treatment plants could apply, however, for EPA permission for industrial users to employ pretreatment standards based on local needs rather than national categorical pretreatment standards. Limitations on thermal discharges, such as those from power generating plants that use surface waters for cooling purposes, could be based on either (1) attainment of thermal water quality criteria, or (2) maintenance of a balanced population of shellfish, fish, and wildlife (rather than on maintenance of an indigenous population of shellfish, fish, and wildlife). The bill would allow thermal dischargers to employ alternative measures, such as the operation of a fish hatchery restocking program, if these measures equal the effect obtained through use of the best technology available.

The bill would extend the date for industry compliance with effluent guidelines to July 1, 1988, and permit achievement of limitations based on best professional judgment within three years of the issuance of these limitations. New source performance standards would only be applied after regulations establishing such standards become final (rather than from the date standards are proposed as is now true). The bill would exempt the discharge of conventional munitions from the coverage of the statute.

EPA's authority to investigate criminal violations of the statute would be confirmed. Criminal sanctions for knowing violations of the statute would be increased to a maximum of $50,000 per day and two years in prison. Introduction of a pollutant or hazardous substance that may cause personal injury or other harm into a sewer system or publicly owned wastewater treatment plant would be made a criminal offense. Civil monetary penalties of up to $10,000 per day of a violation, not to exceed $75,000 in the aggregate, could be assessed by EPA administratively. Violators could obtain judicial review of these assessments in U.S. district courts.

The bill would authorize EPA to charge fees for processing waiver and exemption applications. It would extend the effective life of discharge permits from five to no more than ten years. EPA could delegate the issuance of permits to states even though only a part of this function is delegated to a particular state.

The current bill does not exempt dams as point sources, since a court ruling that dams are point sources covered by the statute has been reversed on appeal. The draft bill does not give EPA waiver authority to excuse compliance with BAT requirements in certain situations, as proposed by some industry groups. No changes would be made in the standards or permit procedures for the deposit of dredged and fill materials in navigable waters and wetlands. The bill makes no change in the current approach of letting the states deal with nonpoint sources of pollution, such as drainage of rain waters across cattle feeding pens.

4

BILLS BEFORE CONGRESS

Bills before the 98th Congress, 1st session, include S. 431, as modified in the Senate Committee on Environment and Public Works. Introduced by Senators John H. Chafee (R-R.I.) and Jennings Randolph (D-W.Va.), it awaits Senate consideration as this is written. That bill and H.R. 3282, introduced by Representative James J. Howard (D-N.J.), are expected to be the primary vehicles for congressional action. References hereafter to S. 431 are to the version approved by the Senate committee on June 28, 1983. Committee agreement may be obtained for other changes.

Other bills include the following: S. 432 introduced by Senator Randolph extends certain compliance deadlines; S. 1288 by Senator Chafee is a one-year reauthorization bill, a shell bill to meet the May 15 deadline for spending authorizations for fiscal 1984; S. 1536 by Senator Christopher H. Dodd tightens the provisions on waiver for discharges from public treatment works into marine waters; H.R. 504 by Representative Bill Chappell (D-Fla.) requires that EPA compliance directives include a cost-benefit comparison; H.R. 754 by Representative Joseph P. Addabbo (D-N.Y.) provides for sludge research and demonstration projects and assistance to local governments wishing to dispose of sludge; H.R. 1570 by Representative Sam B. Hall, Jr. (D-Tex.), restricts federal jurisdiction for the regulation of the deposit of dredged and fill materials in navigable waters and wetlands; H.R. 2190 by Representative Marilyn Lloyd Bouquard (D-Tenn.) deals with local pretreatment programs; and H.R. 2305 by Representative James O. Oberstar (D-Minn.) extends certain compliance dates.

S. 431 as reported in the Senate provides reauthorization authority through fiscal 1987, and H.R. 3282 through fiscal 1988. Neither bill would include the administration's local option pretreatment approach and neither would relax thermal pollution criteria as proposed by the administration.

S. 431 extends compliance deadlines for achieving technology-based standards to three years after promulgation of the standards or three years after July 1, 1983, whichever is later. Dischargers subjected to best engineering judgment permits must be in compliance within three years of the date of the issuance of the permits. All technology-based standards must be met no later than July 1, 1987. H.R. 3282 requires compliance within three years after technology-based standards are established.

Both bills require identification of waters that will not meet water quality standards, notwithstanding compliance with BAT, because of toxic pollutants.

The bills require special measures, including state specification of numerical criteria and the modification of permits of point source dischargers to deal with problem pollutants.

Both bills require states to maintain previously published designated uses for bodies of water within their boundaries, unless, in limited circumstances, these uses are unattainable. They prohibit the degradation of water quality that exceeds the standards required for designated uses. A carefully circumscribed exception would be permitted, however, for essential economic or social development for which there is no feasible alternative. No degradation would be permitted for high-quality waters that constitute an outstanding national resource, such as waters of certain parks and refuges.

S. 431 would allow EPA, with the concurrence of the state, to issue a permit that modifies effluent limits for a particular discharger for nontoxic pollutants on a showing that there is no reasonable relationship between economic and social costs and the benefits to be obtained from the previously set limits. A similar modification, not to exceed five years, could be made for a discharger of toxic pollutants who demonstrates that the modified requirement represents the maximum control attainable within the economic capability of the owner and operator of the facility. The discharger, however, would have to meet certain "reasonable progress" requirements.

H.R. 3282 would increase the maximum civil monetary penalty for violations of the statute from $10,000 to $20,000 per day. These penalties would be assessed in court proceedings. S. 431 would increase the maximum civil penalty to $25,000 per day and permit EPA to assess civil penalties administratively for amounts up to $10,000 per day of violation if the aggregate of such penalties does not exceed $75,000. Violators could obtain court review of administratively imposed penalties in U.S. district courts.

Criminal sanctions are increased and criminal penalties added to cover the introduction of pollutants or hazardous substances into a sewer or public treatment plant if the discharger could foresee that the discharge might cause personal injury or property damage, or might violate an effluent limitation by a treatment plant or a permit condition. S. 431 would allow courts to award attorneys fees and expert witness fees to the prevailing or substantially prevailing party in civil litigation under the statute. S. 431 would also allow private litigants to sue under state court-made law, under other federal statutes, or under the provisions of the Clean Water Act.

S. 431 would authorize EPA to delegate part of its authority to issue permits to dischargers to states unable to assume full responsibility for this function. Both bills would limit the circumstances in which municipalities could obtain waivers of the requirement of secondary treatment to discharge wastewaters into marine waters. H.R. 3282 would delete current authority for waivers for discharges into estuarine waters. S. 431 would only permit discharges into waters whose characteristics ensure appropriate dilution of wastes. Discharges could not be made into saline estuarine waters that do not support a balanced indigenous population of shellfish, fish, and wildlife, do not allow recreation in and on the waters, or do not meet applicable water quality standards.

In the case of toxic pollutants, H.R. 3282 would require that pretreatment and treatment in a municipal treatment plant serving a population of 50,000 or more must remove such pollutants as effectively as secondary treatment in municipal wastewater plants. S. 431 requires that the municipal plant enforce pretreatment standards for toxic pollutants and that those plants serving a population of 5,000 or more must, in the absence of categorical standards, enforce comparable standards.

H.R. 3282 would authorize EPA to convene conferences of states adjacent to an estuary if the action of more than one state is required to attain or maintain clean water standards. The conference would develop a master plan and grants would be authorized equal to 55 percent of the states' implementation costs for the plan. Master plans would deal with both point and nonpoint sources of pollution. S. 431, on the other hand, would authorize a special program to coordinate efforts to clean up pollution in the Chesapeake Bay and to develop a master plan for Narragansett Bay. Grants would be authorized for states adjacent to these bays to implement management plans. A provision of S. 431 approved in the Environmental Pollution Subcommittee that would authorize grants to states presenting plans for the control of nonpoint source pollution was deleted in the full committee. S. 431 would bar EPA from requiring permits for certain discharges of storm water runoff from mining operations and from oil or gas exploration, production, or processing operations.

S. 431 establishes a time frame for EPA regulations dealing with toxic pollutants in sludge. A discharge permit would be authorized for ten years subject to earlier adjustment if new or modified requirements are imposed under certain sections of the statute. A permit would be good for only five years if it includes a waiver or modification. Provision is made for a state or municipality whose water quality is adversely affected by pollutants from another state to petition EPA for an administrative order to resolve the dispute over such pollution.

Only under certain conditions would H.R. 3282 authorize a permit for depositing dredged and fill materials into the waters of the territorial sea, the contiguous zone, or the ocean: if the materials, alone or in combination with other pollutants, would not have an unacceptable adverse effect on the aquatic environment; if there is no practicable alternative to the proposed deposit (taking into account economic and social costs and technological feasibility) with less adverse effect; and if there are no other significant adverse consequences. The bill also spells out new application and approval procedures for permits for the deposit of dredged and fill materials.

H.R. 3282 requires EPA to monitor and study the effects of impoundment and discharge of waters by dams and to submit its control recommendations to the Congress. S. 431 requires that EPA report to Congress on certain hazardous wastes that are exempted from regulation under the Solid Waste Disposal Act because they are waste mixtures that pass into publicly owned wastewater treatment plants.

9

5

PRETREATMENT REQUIREMENTS

The Clean Water Act requires that owners and operators of industrial facilities discharging wastes into publicly owned wastewater treatment plants meet pretreatment standards for all pollutants that are not susceptible to treatment in such plants or that interfere with, pass through, or otherwise are incompatible with such plants. Standards include prohibited discharge standards for pollutants that may interfere with plant operation or performance, and categorical pretreatment standards based on the best available technology that is economically achievable. Contamination of sludge is one of the criteria for requiring pretreatment. The pretreatment program applies to 129 priority pollutants in each of thirty-four industrial categories, but standards have not been promulgated for a majority of the industrial categories. Pretreatment regulations issued by EPA require industrial plants to meet numerical limitations similar to those imposed on direct discharges into surface waters, subject only to a credit to the extent that municipal systems consistently remove a portion of the pollutant before its ultimate discharge into surface waters.

The administration's proposal would allow municipal treatment works to apply on behalf of industrial users for the use of pretreatment standards based on local conditions rather than follow national categorical standards.

Advocates of changes in the existing pretreatment program contend that the program should be modified because categorical standards are not necessary for all industrial categories. The current pretreatment program, when fully implemented, would require a massive commitment to expensive technology-based measures. Deputy Administrator John W. Hernandez, Jr., of EPA testified that well-operated biological treatment systems of municipalities can effectively control many of the toxic materials that concerned the authors of this program. Effluent guidelines based on best practical technology for control of conventional pollutants also effectively control toxic organic pollutants and heavy metals, thus making additional measures unnecessary. Not all priority pollutants present health or environmental risks, moreover, at the levels of concentration at which they are found in wastewater systems.[2] Pretreatment, according to the assistant chief engineer of the Sanitary District of Los Angeles County, requires that firms remove substances that can be more effectively removed in public treatment plants.[3] Representative Robert A. Young (D-Mo.) states that mandating across-the-board national standards for the pretreatment of pollutants requires costly industrial treatment where water quality standards are already being met. This, according to Representative Young, is a massive waste of our capital resources.[4]

Industry critics of the pretreatment program contend that toxic pollutants found in municipal treatment plants may be from sources other than business users of these facilities. A Department of Commerce study found that storm water runoff contributed more of the metals employed in electroplating, other than nickel, than the electroplaters did and that copper sulphate used by the city of New York as an algicide was the major contributor of copper entering sewer plants. Local treatment plant officials criticize the pretreatment program as unworkable, needlessly complex, confusing, and more costly than its environmental benefits.[5]

Pretreatment requirements, critics claim, involve duplicate EPA controls over the same waste materials—controls when wastes enter a publicly owned treatment plant and controls over the level of pollutants discharged by those plants. They note EPA advice that the current pretreatment program will apply to 129 priority pollutants in each of thirty-four industrial categories. EPA has estimated that at least 60,000 industrial dischargers will be required to pretreat wastes and that additional industrial categories and additional pollutants may be added that could affect other dischargers. The Association of State and Interstate Water Pollution Control Administrators estimates that 260,000 to 350,000 firms may have to pretreat wastes if the statute is not changed.[6] Pretreatment imposes particularly high costs on small users who lack the size to enjoy economies of scale. The General Accounting Office (GAO) noted that the program may impose costly, inequitable, and unnecessary requirements and thus use scarce federal, state, and local pollution control resources that could be better spent on other objectives.[7]

The Subcommittee on Oversight and Review of the House Committee on Public Works and Transportation recommended eliminating national pretreatment standards noting that EPA, eight years after enactment of the pretreatment requirement, has been almost totally unsuccessful in carrying out the pretreatment program. The statute should be amended, according to the subcommittee, to permit local treatment plants to establish pretreatment requirements on the basis of their discharge permits.[8] The legislative changes proposed, it is claimed, will allow EPA to concentrate on standards for those industries whose incompatible pollutants deserve this special pretreatment attention. Pretreatment standards as implemented by EPA are a special burden on municipal wastewater treatment plants because EPA regulations require that these plants allow industries credit for the amount of primary pollutants removed by municipal systems. Municipalities, according to GAO, are burdened with the difficulty of computing these credits and prefer not to allow them. Failure to allow credits under the present program, however, would cause industries to engage in unnecessary pretreatment since the categorical standards would continue to require this.[9]

Advocates of the current pretreatment requirements contend that, as noted in the 1979 annual report of the Council on Environmental Quality (CEQ), industrial discharges that have not been pretreated can cause serious problems in municipal plants as many industrial wastes are incompatible with normal sewage treatment processes. Toxic substances and heavy metals, it is claimed, can accumulate in sewage sludge, be volatized during aeration, or pass through the plant untreated. Their presence in sludge can make it unusable for fertilizer. And toxic substances

11

may leach into surface or ground waters if sludge is placed in a landfill.

Toxic substances in industrial wastes, it is argued, may kill bacteria used to treat municipal wastes, thus impairing the treatment capability of the plant. Counsel for the Natural Resources Defense Council contends that curtailing the pretreatment program would permit millions of pounds of toxic chemicals to be released each year.[10]

Critics of the proposal to allow local treatment plants to set pretreatment requirements based on local conditions contend that EPA data show that one-fourth of all municipal plants have interference and upset problems attributable to industrial wastes. Nearly all of these plants, according to a recent study, reported sludge contamination from industrial discharges. Only one-third of the plants monitor for incoming metals or toxic organic compounds.[11]

The National Resources Defense Council (NRDC) estimates that more than half of the toxic pollutants entering public treatment plants come from industrial sources. Many municipal plants still have not installed the rigid biological treatment mandated by the statute. Thus, these plants will not achieve estimated reductions in toxic pollution for many years. The current pretreatment program, according to NRDC, attempts to place a major share of the cleanup burden on the industries that generate the pollutants. Cleanup costs will be passed on by these industries to consumers through increased prices. Consumer resistance to price increases, according to NRDC, will stimulate the industries involved to reduce the volume of their pollutants. Up to three-fourths of the removal costs for these pollutants, if we rely on municipal treatment plants to remove them, are borne by taxpayers, according to NRDC, and there is then little incentive for industries to reduce the amount of pollutants discharged by their plants.[12]

Eric Eidsness, as EPA assistant administrator for water, responds to claims that using standards based on local conditions for pretreatment might not do the job by noting that EPA can withdraw certification for plants that fail to meet requirements.[13]

6

REGULATION OF THERMAL DISCHARGES

The current statute requires that effluent limitations for the thermal component of any discharge into surface waters ensure the protection and propagation of a balanced indigenous population of shellfish, fish, and wildlife in and on that water. The administration proposes that dischargers meet either thermal water criteria or be subject to a test based on balanced populations of fish, etc. Current law requires that cooling water intake structures meet the best available technology standard for location, design, construction, and capacity to minimize adverse environmental impact. The administration would allow alternative measures, such as the construction and operation of a fish hatchery restocking program, to mitigate adverse environmental effects of intake structures, if the alternative measures are equal in effect to the best available technology for intake structures.

The administration argues for these recommended changes because the present standard may require protection for uses that do not exist in the receiving waters and that may exceed existing state standards. Use of the alternative standard would be conditioned on the establishment of a monitoring program to assure compliance. The alternative measures for intake structures would obviate some siting problems for utilities and other industrial plants while preserving the environmental values protected by the current statute. Industry advocates of these changes note that the nation will suffer serious brownouts if additional power generating facilities cannot be sited because of current regulatory restrictions. Compliance with thermal water criteria would simplify monitoring for compliance and thus reduce costs that must be passed on to the consuming public. The changes would permit orderly growth in the national economy without sacrificing environmental values.

Environmentalists argue that energy self-sufficiency should be achieved through conservation and the use of solar power. They fear that allowing alternative measures in lieu of best available technology for intake structures may compromise the attainment of the broad environmental goals they advocate. If Congress should err, environmentalists claim, it should err on the side of protecting fish and wildlife in their native habitats rather than sanction their replacement through restocking.

Advocates of changes in the statute respond that neither conservation strategies nor greater use of solar power has the potential to solve an energy shortfall.

7

WAIVERS FOR MARINE DISCHARGES

The 1972 amendments of the Clean Water Act required that publicly owned wastewater treatment plants achieve effluent limitations based on secondary treatment. They were required to achieve best practicable waste treatment technology by 1983, a deadline extended to 1988 by recent legislation. Amendments adopted in 1977 permitted waivers of the requirement of secondary treatment for municipal treatment plants that were already discharging wastes in marine waters. Because of delays in issuing regulations, the restrictive conditions imposed, and the limited number of municipalities that could qualify, few municipalities were able to submit timely applications for waiver.[14]

Congress adopted legislation in December 1981 that reopened the application process for marine discharge waivers. S. 431, S. 1536, and H.R. 3282 all seek to establish more restrictive conditions for the granting of these waivers. S. 1536, for example, would require full primary treatment of effluents, mandate the equivalent of secondary treatment for toxic pollutants, and limit waivers to discharges in deep, well-flushed waters. Waivers would be denied if collective discharges of several treatment plants under such waivers could degrade water quality or if water is already polluted from other sources. Senator Dodd argues that the 1981 amendments to the statute stimulated a flood of waiver applications. He fears that the granting of these applications may degrade many coastal waters and thereby injure commercial shellfishing and recreation.[15]

Those who oppose more restrictive conditions for waivers for marine discharges contend that current law provides adequate protection through provisions that require the attainment and maintenance of ambient water quality, that monitor discharges, and that restrict discharges from increasing above the volume specified in the waiver. A study published by the GAO in 1981 identified potential savings of up to $10 billion if waivers were allowed for 800 communities of modest size. These savings, according to GAO, are what communities would realize from not building secondary treatment facilities; they do not include the costs of operation and maintenance that would also be saved.[16]

8

DEPOSIT OF DREDGED AND FILL MATERIALS

The administration has not recommended legislative changes in the current permit process or in standards for the deposit of dredged and fill materials into navigable waters including the territorial seas and wetlands in contact with the seas. H.R. 3282 would tighten the criteria considered in evaluating deposit applications and spell out procedures for processing applications that would ensure consideration of the views of EPA and other agencies and the right of these agencies to appeal to higher authorities within the Army Corps of Engineers.

The statute currently entrusts the granting of permits, after public notice and an opportunity for a public hearing, to the secretary of the army. This authority may be delegated to the states that are willing to qualify under detailed requirements set forth in the statute. Certain activities, including projects specifically authorized by the Congress and normal farming, silviculture, and ranching operations, are excepted from the permit procedures. These exceptions would not be rescinded by H.R. 3282. EPA, of course, is responsible for developing environmental guidelines that the secretary of the army (acting through the Corps of Engineers) must observe in considering permit applications. EPA may prohibit the designation of any disposal site if its use would have an unacceptable adverse effect on municipal water supplies, shellfish beds, fishery areas, and wildlife or recreation areas.

Advocates of tighter restrictions on the deposit of dredged and fill materials contend these restrictions are needed to force the Corps of Engineers to take the advice of EPA and other agencies seriously. Such restrictions would help maintain fishable and swimmable waters and protect water supply sources. Advocates contend that wetlands protection is required because liberal deposits of dredged and fill materials may cause serious and permanent ecological damage. The wetlands, bays, estuaries, and deltas that are protected by the permit procedures are among the most biologically active areas in the country. They are the spawning grounds for fish and shellfish, the nesting sites for birds, and a major situs of food for wildlife and fish.

Advocates of limiting federal jurisdiction over the deposit of dredged and fill materials maintain that too many minor activities are subject to intensive regulatory review under the current statute. Applicants in 1980 sought over 10,000 permits; over 8,000 permits were in fact granted. The process for denying or approving these applications, however, is far too slow. With the confusion of overlapping agency jurisdiction and as many as four layers of review, agencies frequently fail

15

to observe agreed time schedules for passing on applications. Action time for controversial applications averages 271 days.[17]

Senator John G. Tower (R-Tex.) argues that federal jurisdiction should be limited to waters that are truly navigable. The government, according to Senator Tower, has used current restrictions to deny property owners the right to deposit dredged and fill materials on their own property. Similarly, citizens, navigation districts, and municipalities may be denied the right to undertake flood control projects.[18] Senator Lloyd M. Bentsen (D-Tex.) points out that current restrictions would preclude the deposit of materials in normally dry depressions in semi-arid regions. There are 10,000 of these "lakes," most of which are privately owned and used for agricultural purposes.[19] Restricting federal jurisdiction would eliminate interference with the deposit of dredged and fill materials except in the truly navigable waters and would end the antidevelopment bent that inhibits economic growth.

9

COST CONSIDERATIONS

The administration has not proposed major changes in the regulatory strategy of the current statute to reduce compliance costs nor has it recommended the use of marginal cost-benefit analysis in setting standards. Economists and others criticize the costly approach of the present statute and cite a variety of alternatives. This chapter will examine criticisms and advantages of the current strategy and the arguments for and against the principal alternatives.

CRITICISMS OF THE CURRENT STRATEGY

Nationally Uniform Technology-Based Standard Strategy. Economists and business persons particularly criticize the uniform technology requirements mandated in the 1972 amendments. Technology-based standards uniformly imposed on industry regardless of local conditions amount to basing standards on what can be done and not on what should be done. Why, critics ask, should BAT be required everywhere when receiving waters differ in quality from one stream or body of water to another and from one stretch of a stream to another? John W. Hernandez, Jr., when deputy EPA administrator, argued that technology-based standards set by EPA are laboratory-derived and cannot possibly reflect stream-specific physical data such as hardness of water, pH, and the resident species of fish actually encountered.[20] Surface waters can assimilate more wastes when there is greater water flow or when temperatures are lower. Phosphorus controls may be appropriate for a lake in danger of eutrophication but not for flowing streams with no problem of high phosphorus.[21] Various segments of a river differ in their capacity to assimilate wastes without harming aquatic life or reducing recreational opportunities. And in some instances controls may be futile anyhow, because an ample supply of nutrients may be available to promote eutrophication regardless of the treatment called for by technology-based standards.[22] Thus, the Subcommittee on Investigations and Oversight of the House Committee on Public Works and Transportation concluded that consideration should be given to the technology required in view of the general characteristics of the receiving stream, the nature of the wastes involved, and the affordability of various available treatment processes.[23]

The standards employed under the clean water legislation are referred to as technology-based because they are tied to standards that are achievable by employing known technology. Critics contend that such standards chill the incentive to discover more effective technologies. Dischargers have only the incentive to meet the required standard and no more. Should they develop technologies that produce water quality benefits exceeding those of the required standard, they only hand EPA regulators the means of imposing a new standard on them and on all those similarly situated at additional cost to industry.[24] Imposing water quality standards, it is claimed, would leave the selection of technology to dischargers; they would then have the incentives necessary to develop the most cost-effective strategies and technologies possible.[25]

The statute's statement of the technology to be employed by dischargers and the qualifications attached, such as "best practicable" and "best available," according to critics, is ambiguous and imprecise. EPA, using these and similar evidences of congressional intent, had to set nearly 30,000 standards in a relatively short period of time. (Separate standards were required for various pollutants for forty-six industries and numerous subcategories per industry.) Small wonder, according to critics, that national standards not tied specifically to local conditions and allowing a "margin of safety" imposed major compliance costs at locations where discharges caused no discernible harm to water quality.[26] The regulatory strategy decreed in the statute requires that, if technology can reduce the amount of pollutants anywhere, the technology-based standards must be applied everywhere. Firms located in certain areas are forced to incur costs that are unnecessary or excessive compared with the benefits obtained. This, critics claim, erodes popular support for the achievement of environmental objectives.[27] Some studies show a potential for cost savings of 30 to 35 percent nationwide from the selection of cost-minimizing patterns of effluent reductions that would achieve the same pattern of water quality improvements that can be obtained with technology-based standards.[28] GAO, critics note, asserts that the need for rigorous nationally uniform technology-based treatment standards in current law is as yet unproven and the public benefits gained from the additional investment required to meet these standards are not apparent.[29]

Defenders of technology-based standards contend that, when water quality standards were employed prior to the 1972 legislation, enforcement officials found it difficult to establish reliable and precise effluent limitations for dischargers that would produce the stream quality desired. Enforcement efforts under the former regime were hindered because enforcement officials had difficulty establishing a precise link between the effluent discharged by a particular plant and the degradation of water quality. Establishing this link is particularly hard if numerous dischargers send wastes into the same body of water. Shifting to the uniform technology-based regulatory approach eliminates the need to search for direct cause and effect between a plant's discharges and lowered water quality. Indeed, dischargers failing to meet required effluent standards can be proceeded against without any proof of the effect of their effluents on water quality.[30] Simplifying the enforcement effort leads, it is claimed, to greater compliance by industry and thus to more rapid attainment of good water quality. The burden on enforcement officials of setting uniform standards is less than the burden of enforcing varying water quality standards, particularly in view of the difficulty of accurate measurement of water quality. Moreover, if dischargers are allowed to minimize costs by making a smaller cleanup effort on effluents where the receiving waters are cleaner, as in the selective cost-minimizing strategy cited above, this leaves no room for additional pollutants in the water if new plants are located in the area. In other words, standards for existing plants would have to be repeatedly revised to make room for new plants.[31]

Defenders of the current strategy argue that substitution of water quality standards would also be disadvantageous because local officials would necessarily have to allocate the cleanup burden among local plants. Plants that might have to incur major cleanup costs would have greater leverage on local officials through threats of plant closings than they would have on national officials. Thus, the cleanup might slow perceptibly or grind to a halt. Switching to water quality standards, it is claimed, would give dischargers a "license to pollute," up to a certain point. The technology-based standards, on the other hand, foster a nondegradation policy for clean surface waters and remove the incentive for polluters to move from older industrial areas to areas with clean water. Thus, the current policy, it is claimed, minimizes the closing of factories in depressed areas where the impact of unemployment would be more serious.

Defenders of uniform standards deny that the current strategy seriously discourages innovation into better and most cost-effective control strategies. Dischargers can comply with the standards using any technologies they want, as long as they achieve the level of control required by EPA standards.

Critics of the current regulatory strategy contend that setting water quality standards would not license pollution but would rather recognize that no extant technology will allow dischargers to achieve a zero discharge rate. Water quality standards offer a more realistic and cost-effective way to deal with pollution, and dischargers are more likely to cooperate if the regulatory regime is reasonable. Such savings as enforcement officials may gain from technology-based standards applied uniformly throughout the United States are more than offset by unnecessary compliance costs imposed on industry. Many other factors outweigh pollution control costs in decision making on plant location, it is claimed, and no major exodus from older industrialized areas can be expected.

The Regulatory and Grant Strategy for Municipal Plants. The 1972 statute replicated its strategy for controlling industrial pollution by adding a "carrot" in the form of federal grant moneys to help fund municipal treatment plants. The statute mandated secondary treatment of wastes as a general proposition, and no less costly standard would suffice.[32] Critics of this program contend that the secondary treatment standard coupled with the ultraconservative criteria in EPA's Red Book require the overdesign of wastewater treatment plants.[33] Towns downstream from major industrial areas have used federal funds to build treatment plants that discharge effluents that are of higher quality than the waters into which the effluent is discharged.[34] Some critics charge that the uniform national standard for these plants produces too much treatment for plants in some locations and too little treatment in other locations. These critics advocate the use of less expensive treatment methods where this is feasible, the use of water quality standards to minimize costs in many locations, and such alternatives as seasonal variations in effluent standards, physical measures to increase the assimilative capacity of water, and the use of a bubble approach as under the Clean Air Act.

The incentive structure inherent in the municipal wastewater plant program, it is claimed, clearly produces the most expensive solution for publicly owned wastewater plants. Federal funding of a major part of construction costs leads· municipalities to opt for grandiose solutions when less lavish plans would do the job at less overall cost. Municipalities do not know how long federal grant funds will be available; their incentive therefore is to build in as much excess capacity as they can to cover future needs and future city development. Kneese and Schultze cite data showing that fully a quarter of the waste treatment capacity of metropolitan area plants was less than half used.[35] The federal subsidy, it is claimed, induces planners of municipal plants to choose expensive plants with high capital costs relative to "on paper" operating and maintenance costs in the belief that the federal government will thus share a higher percentage of the life cycle costs of the facilities.[36] The incentive to build extremely costly facilities is accentuated, some critics claim, because engineering firms that design treatment plants generally charge a percentage of total construction costs for their services. The more expensive the plant the larger the firms' fees.

Since operating costs and maintenance are not federally subsidized, municipalities, which must fund these expenses in their entirety, often attempt to economize with fewer staff members than desirable, with less-experienced personnel, and with inadequate maintenance. EPA approval of grant applications after review does not, it is asserted, guarantee that the completed facilities will work properly; indeed many do not work properly.[37] EPA's own data, it is claimed, show that as many as 50 percent of these new facilities fail to operate properly because of improper system design and installation, poor training of operators, and inability in some cases to attract qualified personnel. The result is that the statute's goals are not met and the need for federal funds grows even larger.[38] A GAO survey of 242 treatment plants found that 87 percent were in violation of EPA standards for at least one month out of twelve, and 56 percent of the violating plants exceeded their discharge permit limits for pollutants for more than six out of twelve months.[39]

The technology for secondary treatment has been known since 1910, critics claim, and fear that innovative technologies may not work has led planners to replicate the same plant designs and technologies, continuing to build the same expensive types of plants. Substituting water quality standards would do more to encourage planning for the right amount of capacity for pollutant removal at the least overall cost than would tinkering with grant formulas while the secondary treatment requirement is left in place. Kneese and Schultze maintain that the most fundamental objection to subsidy arrangements is that they do nothing to mitigate the excess generation of wastes in the first place. They do not lessen the incentives to use waterways for sewage disposal, and they ignore all methods for managing water quality other than conventional treatment of sewer wastes.[40] An optimal basin-wide plan would relate the desired level of treatment to stream-flow conditions and downstream uses, among other variables, at substantial savings to taxpayers.[41]

Defenders of the current strategy for control of pollutants from municipal treatment facilities contend that uniform standards avoid the very difficult task of determining effluent levels that will achieve a certain quality of water in receiving streams and lakes. The strategy parallels that employed for industrial dischargers, it is claimed, except for the assistance provided for the building of acceptable wastewater treatment plants. The grants are the carrot the drafters used to induce compliance within the time frame set by the statute. Even though there have been problems with the projects constructed under the grant program, advocates insist that grants are wanted, and needed, if we are to attain clean water goals. Corrective measures have been taken. The federal grant share has been cut to 55 percent of future projects to remove the incentive to build costly plants. The 1981 amendments permit the use of biological treatment facilities as the equivalent of secondary treatment in appropriate circumstances, facilities that are often cheaper and more energy efficient.

The statute has been adjusted to encourage priority funding for the projects that will provide the greatest water quality benefit. Lower reserve capacity for growth will be permitted for future projects, and the prime consultant to a municipality constructing a new plant will be required to remain on the site for a year after plant startup to ensure that it works satisfactorily and in compliance with the municipality's permit.[42]

PROGRESS TOWARD CLEAN WATER

Critics of the strategies employed in clean water legislation contend that the program is not cost effective and that it has failed to achieve satisfactory water quality. The goal of fishable-swimmable waters by July 1, 1983, has not been met, and the elimination of all discharges into our surface waters by 1985 is equally unattainable. Critics cite the report of the Council on Environmental Quality based on U.S. Geological Survey data and EPA water quality definitions to show that, while there has been substantial improvement in water quality in isolated instances, the quality of surface waters nationally did not change much between 1975 and 1979.[43] While about 81 percent of industrial dischargers met the 1977 deadline for best practicable control technology, only 42 percent of the publicly owned facilities met secondary treatment standards by that date.

The clean water legislation primarily emphasizes control of indentifiable point source discharges of pollutants, that is, discharges from conduits as distinct from diffuse sources such as runoff of surface waters after a rain and airborne pollution. Critics of the point source strategy contend that not enough attention has been paid to nonpoint source pollution. GAO, in fact, concludes that progress in controlling nonpoint source pollution is minimal.[44] An EPA staff report places the nationwide average of pollutants from nonpoint sources at 92 percent for suspended solids, 98 percent for fecal coliforms, 79 percent for total nitrates, 53 percent for total phosphates, and 37 percent for biochemical oxygen demand (BOD).[45]

A survey in Chicago is cited to show that, if all point sources for cadmium, for example, were controlled to zero discharge levels, the cadmium content in the city's sludge would be reduced only 40 percent.[46] The comptroller general testified in 1979 that the goal of fishable-swimmable waters could not be met unless more was done to control pollutants from nonpoint sources.[47] Thus, according to critics, the current point source strategy of clamping down harder on identifiable point sources while not doing enough to deal with pollution from other sources not only is very costly, but also is unlikely to achieve the goals of the legislation even over a longer period of time. When, critics wonder, will the massive federal grant program ever end as water quality benefits from these expenditures seem to be more than offset by continuing problems with plant performance and with uncontrolled nonpoint source pollution?[48] Critics fear that clean water goals cannot be met with current strategies regardless of cost to industry and taxpayers.

Defenders of the strategies employed in the 1972 legislation maintain that the 1977 compliance rate for best practicable technology among large industrial dischargers was a major accomplishment. The nation's resources are finite, and this includes the assimilative capacity of its surface waters. Given these limitations and the nation's population growth, the fact that our surface waters are not deteriorating is, in the view of CEQ, a victory for the nation's water pollution control effort.[49] Defenders cite the testimony of Hernandez that implementing BPT requirements significantly reduced industrial discharges of six key pollutants from 1972 to 1977: BOD by 69 percent; total suspended solids (TSS) by 80 percent; oil and grease by 71 percent; dissolved solids by 52 percent; phosphate by 74 percent; and heavy metals by 75 percent.[50] Greater gains were not made from the grant and regulatory program for publicly owned wastewater treatment plants, it is claimed, because only 2,600 of the planned 19,000 projects funded with the aid of federal grants are completed and operational. By 1980 these plants were removing 65 percent more BOD and TSS than was being removed in 1973 and by the year 2000 it is claimed there will be a 101 percent increase in BOD and TSS removal over 1973 figures.[51]

Environmentalists defend the statute's concentration on point source polluters. The greatest gains can be made in the least time by singling out these identifiable dischargers. Direct restrictions on point source polluters, it is argued, eases enforcement officials' work significantly and, without this change in 1972, environmentalists feel even less progress would have been made. Strong point source regulation is important, it is claimed, because the projected growth of organic wastes from municipal use of water may reach 300 percent between 1970 and the year 2020. Ever larger amounts of organic waste are generated by industry. The food processing, chemical, and paper industries are major sources of organic waste. These industries contribute 90 percent of all organic wastes from manufacturing.[52] Thus, environmentalists view the current strategies as correct and needed to cope with a growing volume of waste. They rely on section 208 of the statute to stimulate areawide planning for control of such nonpoint source pollution as agricultural and construction runoff and surface and underground mine drainage.[53]

22

Critics of the clean water program and some proponents are concerned about the high cost of the program. When the 1972 statute was enacted, the Senate committee report cited EPA's estimate that $12.6 billion was needed to provide secondary treatment, and some additional treatment where a need for such treatment is identified, by all communities with sewer systems. A survey of the U.S. Conference of Mayors cited in that report estimated a plant construction backlog of $33–$37 billion.[54] That report contained no estimate of the costs to industry. Actual expenditures for clean water regulatory requirements through 1981 are shown in table 1.

TABLE 1

CLEAN WATER EXPENDITURES FOR 1972–1981
(millions of current dollars)

Year	Government Spending on Pollution Abatement	Business Spending on Pollution Abatement	Total Spending (Including R & D, Government Regulation, & Monitoring)
1972	$ 3,483	$ 4,957	$ 8,727
1973	3,961	5,770	10,089
1974	4,881	6,354	11,634
1975	5,768	7,349	13,561
1976	6,353	8,653	15,513
1977	6,406	9,862	16,838
1978	8,038	11,215	19,876
1979	8,681	12,446	21,799
1980	8,511	13,163	22,424
1981	7,128	13,844	21,724
Total	$63,210	$93,613	$162,185

NOTE: Total spending column exceeds sum of first two columns because it includes the costs of regulating and monitoring at national, state, and local levels, as well as government and private R & D expenditures.
SOURCE: For 1972–1977, Department of Commerce, *Survey of Current Business*, vol. 62 (February 1982), table 1 at pages 51–53; for 1978–1981, Department of Commerce, *Survey of Current Business*, vol. 63 (February 1983), table 1 at pages 15–17.

The extent of underestimation of costs of the cleanup and maintenance of surface water quality requires consideration of projected future spending as well as past expenditures.[55] In its 1980 report CEQ projected expenditures by governmental bodies for the period 1979 to 1988 (a three-year overlap with figures cited in table 1) at $48.9 billion in capital costs and $56 billion in operations and maintenance costs. Business capital costs are estimated at $99.7 billion and business operations and maintenance costs at $45.4 billion. The aggregate additional cost in

1979 dollars by this projection is $250 billion.[56] The 1980 needs survey conducted by EPA identified $120 billion in treatment plant construction projects eligible for federal matching contributions to meet needs through the year 2000. Other categories of need not eligible for federal matching funds would raise the aggregate to $233 billion, and some past estimates have run as high as $500 billion.[57] These estimates of need relate solely to governmental spending for treatment of wastes and not to business needs.

There are other costs of the clean water program. Efforts to meet wastewater treatment standards may force the disposition of some wastes into environmentally sensitive media other than surface waters, for example, the burning of sludge, where air rather than water would have to assimilate residuals of this process. GAO points out that municipalities meeting rigorous clean water standards face the unanticipated difficulty of generating more sludge than they would at lower levels of wastewater treatment. The sludge problem has not been satisfactorily resolved. Ocean dumping has been banned to protect surface waters; land fill, composting, and thermal dehydration pose special environmental problems; and the burning of sludge, assuming air quality standards are not violated, may be far too costly, particularly in periods of fuel shortages.[58]

A cost of waste treatment that is frequently underestimated is that of the expense of operating and maintaining a treatment plant once it is built. Martin Long, former president of the Water Pollution Control Federation, notes: "There is a big difference between a citizen's being asked to affirm his support for clean water and telling him his sewer rates have just tripled."[59] CEQ cites an EPA survey of 258 facilities for communities under 50,000 in population and found that operating costs could exceed $100 per household per year in 40 percent of the towns surveyed and over $200 per year per household in 10 percent of the towns. Projected annual costs range near $300 per household per year for communities with fewer than 10,000 people. Many small town householders, as CEQ notes, have incomes well below the national median and thus have a harder time meeting higher costs of waste treatment. Technical requirements of clean water regulations may be unrealistic in the case of Alaskan natives, Indian tribes, and rural communities.[60]

Critics of the business spending data cited in table 1 contend that perhaps as much as 25 percent of business capital expenditures may involve improvements in manufacturing processes. Since these firms derive other benefits from these expenditures, critics claim that such expenditures should be divided between environmental cleanup and regular business investment and that the business spending total should be adjusted downward accordingly.

Businesses contend, on the other hand, that end-of-the-pipe treatment to achieve similar reductions in pollutants might require just as much capital expenditure. Required reductions in discharged pollutants could not be achieved in some instances, it is claimed, by end-of-the-pipe treatment, and thus the capital spending data are very close to the mark.

Some environmentalists contend that comparing costs and benefits is not appropriate in the case of the clean water legislation. Not only do costs differ from industry to industry but the benefits of waste reduction vary by geographic area, type of effluent, timing of discharge, and so forth. Some costs can be fixed with a reasonable degree of certainty, but placing dollar values on benefits is more difficult. How, these environmentalists ask, can one place a value on life generally or on the aesthetic benefits of clean surface waters? Other environmentalists, sensitive to the criticism of the cost of the nation's cleanup effort, are willing to estimate benefits from cleaner water. Economic and health benefits of cleaner water, they say, have been underestimated.[61] And it is difficult to evaluate benefits. There are many indicators of water quality, and they take on different values depending on the time and location of water samplings. The effect of organic wastes on water is complicated. Water quality in the same body of water varies over space and time. There are many different bodies of water nationwide with differing characteristics that make generalization of overall water quality from samplings in certain bodies inaccurate.

Space does not permit a detailed critique of the various water benefit studies that have been made and of the methodology employed. The methodology depends, of course, on the water uses involved, that is, on whether the surface water is to be used, for example, for drinking water, for water contact sports, or for boating and fishing.[62] A synthesis of certain estimates of water quality benefits from the clean water legislation was performed for CEQ by A. Myrick Freeman III.[63] Even though Freeman's synthesis is limited to the water quality program's effect on conventional pollutants, assuming that 1985 statutory requirements for those pollutants are met, and ignores nonpoint sources of pollution, its estimates usefully demonstrate the types of values set for particular uses. Recreational uses of surface waters predominate among the beneficial uses designated for such waters by state authorities. Thus, Freeman's benefit estimates and the estimates of others place a greater value on recreational uses than all other uses combined.

Freeman estimates improved water recreational opportunities (assuming 1985 compliance with the clean water legislation) at $6.7 billion per year, including marine sports fishing at $2 billion, boating at $2 billion, fresh water sports fishing at $1 billion, swimming at $500 million, and waterfowl hunting at $200 million. Improvement in the appearance, taste, and odor of surface waters represents utility gains; if people are willing to pay for these improvements, monetary values can be placed on these characteristics.[64] Freeman estimates improvement in aesthetics and property values at $2 billion. From savings in days lost from work, in medical treatment, and in lives (as they would be valued in court suits), Freeman estimated drinking water and health benefits from cleaner water at $1 billion. Savings in treatment costs for municipal water supplies taken from surface waters are estimated by Freeman at $900 million.

If polluted waters reduce or eliminate a spawning run, eliminate crabs or lobsters, or make species unfit to eat, an economic loss can be established through market value of increased yield net of expenses of other variable factors of production. Freeman estimated improvements in commercial fishing from cleaner water at $800 million. He assigns $600 million as a saving in water treatment costs for industry. Families may incur costs due to the hardness of water used in the home

25

and the presence of dissolved solids in the water. These may produce mineral deposits, may stain fixtures, and may damage appliances. Freeman assigns a value of $300 million to cleaner water for home use. Freeman did not, as some scholars do, place a value on improvements to navigation. Damage from unclean water may involve acid corrosion to vessels and damage to docks, accidents due to floating materials, and losses from water route closures.

Some environmentalists contend that there are also gains in employment from the construction of wastewater treatment plants and the like. EPA has claimed that $4 billion in grant outlays by the federal government together with state matching funds of 25 percent will generate 176,000 work years of additional work. Environmentalists insist that any tally of benefits from the program for cleaner water should include increased interim employment.[65] Environmentalists also claim that new job opportunities and increased national output occur in research and in the operation and maintenance of water treatment facilities.

Economists and business persons on the other hand, note not only that employment opportunities in construction of wastewater plants are temporary but also that during periods of relatively full employment there is no basis for claiming secondary benefits of additional employment. This is true, it is asserted, because this construction requires that other activities be forgone or reduced. Thus the benefits claimed are offset by the loss of benefits from other activities that must be put aside.[66] Indeed, business persons assert, clean water controls raise the cost of doing business and, if these costs cannot be passed on to the consuming public, the business cannot continue to operate and employees will be thrown out of work. When the firm affected is the primary local employer—a cannery or beet sugar factory in a small western town or a pulp mill in a small New England town—the resulting unemployment is particularly painful. Smaller businesses, it is claimed, are those hardest hit by pollution control expenditures and the ones least able to meet enforcement demands and to pass on costs to consumers.[67]

Critics of water quality benefit estimates contend that environmentalists and those retained by them to perform estimates too often adopt a "more environmental than thou" approach and overestimate benefits.[68] They cite, for example, the fact that Freeman, who placed a $1 billion value on clean water benefits to drinking water and health had stated in another publication that it was not clear that water pollution has affected health or mortality in a quantitatively significant way.[69] It is one thing to estimate demand for facilities but quite another to make accurate surveys of potential users and to be sure that the demand is effective demand, that is, that potential users are serious about recreational facilities, would be willing to pay the cost of improved water quality, and have the ability to pay.[70]

Business persons also claim that pollution control expenditures divert resources from production and reduce the productiveness per unit of investment, thus increasing product prices and intensifying inflation.[71]

Environmentalists deny that clean water regulations have more than a nominal effect on productivity and inflation.[72] Proximity to markets, availability of transportation, labor costs, financial problems, and political stability are claimed

26

to be more influential than environmental regulations in determining whether a plant will be closed and relocated abroad or elsewhere.[73]

MARGINAL COST-BENEFIT ANALYSIS

Critics of the 1972 clean water legislation contend that Congress failed to give adequate attention to balancing costs and benefits when it adopted the legislation and that marginal cost-benefit analysis should be employed in setting standards and goals under the statute. The statute, in their view, should be amended to require such analysis, and an evaluation of the benefits of alternative uses of our resources. Only by using cost-benefit analysis can decision makers judge the true merits of environmental regulations and balance those requirements with other national priorities, such as increased productivity, a healthy economy, and adequate energy resources.[74] We cannot have everything. Water cleanup policy, it is claimed, has proceeded on the assumption that we have infinite resources and can in fact have more of everything. Tradeoffs, however, are necessary between environmental quality objectives and other national objectives. Using more resources than are absolutely necessary to achieve pollution control objectives is wasteful. Society should be economical about its choices of environmental goals and standards. We can make optimum choices regarding pollution control only after comparing the marginal cost of increased control with the benefits that will be gained from those expenditures at the margin.[75]

There is no evidence, it is asserted, that costs of achieving clean water were fully considered in the context of the marginal cost-benefit discipline. Had that discipline been applied, far different strategies and levels of control might have been adopted. Nor, critics claim, is there any evidence that EPA employed this discipline in setting water quality guidelines and standards or in calculating the marginal treatment costs to subcategories of industry. Marginal cost-benefit analysis is an attempt to replicate for the public sector the decision that would be made if private market discipline could be applied.[76]

The clean water statute set the goal of fishable-swimmable waters by July 1, 1983, and the goal of complete elimination of all discharges of pollutants into surface waters by 1985. We cannot, however, make the environment pristine, reducing all environmental risks to zero. Professor Lester Lave contends that we need to redefine our environmental goals. Recognizing that health risks from pollution seldom constitute an emergency, Lave argues for quantification of risks in evaluating environmental programs. He estimates the risk of increased lifetime incidence of cancer from the wastes at Love Canal at perhaps 1 percent. Life style, including smoking habits and diet, have much more effect on disease rates, including cancer, according to Lave, than current exposure to environment.[77]

Other scholars point out that as more and more pollutants are removed from effluents, the marginal cost of each additional increment of removal becomes greater and greater. A 1972 study of the cost of removing BOD from water discharges at a typical large meat processing plant illustrates the principle. When

30 percent of BOD had been removed, the cost of removing an additional pound of BOD was 6 cents; but once 90 percent of BOD had been removed, the removal of another pound cost 60 cents; above the 95 percent removal mark the removal of each additional pound costs 90 cents.[78] Similar cost patterns have been demonstrated for other pollutants. The steep rise in incremental removal costs means that moving from 97 to 99 percent removal may cost as much as the entire effort to remove the first 97 percent of pollutants. The last few percentage points of pollution removal can involve huge costs, and these costs represent the value of resources that will not be available for meeting other wants of society. The zero discharge goal it is claimed, will be extremely expensive for all industries to achieve and impossible for some.[79]

Costs associated with various levels of control must be calculated and compared with the benefits of control. Table 2 illustrates estimated costs and benefits at various levels of control and, in the last two columns, marginal costs and marginal benefits.[80]

TABLE 2

COST-BENEFIT ANALYSIS OF DELAWARE ESTUARY COMPREHENSIVE
POLLUTION STUDY CONTROL PLANS
(estimated costs and benefits in millions of dollars)

Quality Level	Total Cost	High and Low Estimates of Benefits	Cost Added over That at Next Lower Level	Added Benefits (High and Low Estimates) over Those at Next Lower Level
I	490	355–155	215	35–20
II	275	320–135	120	10–10
III	155	310–125	45	30–10
IV	110	280–115	80	280–115
V	30	—		

SOURCE: Adapted from the table at p. 187 in Aaron Wildavsky, "Economy and Environment/ Rationality and Ritual," *Stanford Law Review,* vol. 29 (November 1976).

As table 2 shows, moving the quality level from IV to III costs more than the estimated benefits of that action, and the cost of each higher level of quality becomes increasingly more expensive than the estimated benefits to be realized. In other words, at higher levels of water quality massive and certain costs, according to cost-benefit advocates, clearly outweigh the small and uncertain benefits that may be realized.

Setting radical goals and unattainable deadlines may have been a strategy to convince dischargers that the federal government was serious about the cleanup of surface waters, according to critics, but the government sustains a serious loss of credibility when these goals are not attained or when their attainment is prohibitively expensive. It is claimed that even the proenvironmental National Commission on Water Quality recognized this danger and recommended that Congress redefine the zero discharge goal to stress conservation and reuse of resources. The commission noted that attaining the 1985 goal would be prohibitively costly and energy-intensive and would create large quantities of residuals that must be disposed of in ways other than by dumping them in the water.[81] Critics of the strategy of the 1972 statute note GAO's conclusion that "the need for the rigorous water quality standards which form the basis for existing programs is as yet unproven and the public benefits to be derived by additonal investment to meet the standards are not apparent."[82]

The administration's proposal eschews marginal cost-benefit analysis. It does provide, however, that states can downgrade designated uses for bodies of water if they can show that the costs of achieving these uses far outweigh the benefits. Representative Chappell's H.R. 504 would require that EPA compliance directives include cost-benefit comparisons. S. 431 would allow EPA, with the concurrence of the state, to issue a permit that modifies effluent limits of a particular discharger for nontoxic pollutants on a showing of no reasonable relationship between economic and social costs and the benefits of the previous limits. A discharger of toxic pollutants who demonstrates that modified requirements represent the maximum control attainable within the "economic capability" of the owner and operator might also be afforded relief.

Aside from the foregoing the pending bills do not call for cost-benefit comparisons or for marginal cost-benefit analysis.

Environmentalists, on the other hand, contend that cost-benefit analysis makes economic efficiency our primary goal, for which the goal of a clean environment would be traded at margin under economic theory. But public and private preferences belong in different categories, and private preferences should not be substituted for public goals.[83] Aside from the real estate market, environmentalists contend that environmental goods can be demanded only through political institutions. Since citizens cannot pay for environmental goods directly and thus provide market signals for public action, political leaders must assess the public's demand for environmental goods. This is what happened in the enactment of the 1972 water pollution control legislation. Congress was not unmindful that substantial costs might be incurred, including a certain number of failed businesses, but it considered clean water to be worth the price.[84]

Deadlines and goals were included in the clean water legislation, it is claimed, to serve as action-forcing mechanisms. Much more progress has been made in cleaning up the nation's waterways under their pressure than would have been possible without them. Since businesses and communities will always find excuses for not meeting goals and deadlines, it is important that very ambitious goals be included in the relevant statute.

Quite aside from the benefits that some decision makers might gain from the results of cost-benefit analysis, environmentalists contend such analysis is not practical because water quality data are too sketchy to permit accurate evaluation. Any balancing of costs and benefits would necessarily be very crude since costs as well as benefits vary widely depending on plant location and other factors.[85] In addition, counsel for the Natural Resources Defense Council insists that cost-benefit analysis is time-consuming, expensive, and resource intensive.[86] Having set our hands to the task of cleaning up surface waters, environmentalists insist, we should not turn back.

The administration's proposal to allow states to downgrade designated uses of surface waters if costs far outweigh benefits is attacked by environmentalists. S. 431 and H.R. 3282 would freeze designated uses so states could not downgrade them. Critics of this feature of the administration's proposal contend that even this form of cost-benefit balancing would interfere with achieving the statute's goal of fishable and swimmable waters and zero discharge of pollutants into those waters. They fear that those wishing to downgrade designated uses could easily perform the cost-benefit comparison that would permit such action.[87]

Critics of attempts to freeze designated uses contend that current designations may be inappropriate. Uses are often vaguely drafted, high background levels of pollution may make designated uses unattainable, and advanced treatment to achieve earlier designations of uses may simply be unaffordable.[88]

ALTERNATIVES TO DIRECT CONTROL OF WATER POLLUTION

Some economists contend that a better way to achieve clean water objectives involves the use of taxes or charges on polluters, or the substitution of marketable permits to pollute, and some contend the same objectives can be attained by the use of subsidies rather than by charges. The objective of these devices is to harness the efficiency of the market to achieve clean water objectives at less cost overall. Such alternatives to the command and control regulation of the clean water legislation will be examined briefly.

Effluent Charges or Taxes. Without some device such as an effluent charge or tax, the market gives the wrong signals to dischargers. The assimilative capacity of surface waters is a free good to them, a free disposal system that, without government intervention, they would overuse.

The command and control regulatory system has serious problems. If the system is simple enough to be handled by a central bureaucracy, as some think is possible with uniform treatment strategy embodied in the 1972 legislation, it is bound to be very inefficient. If, on the other hand, the system seeks to accommodate the tremendously diverse conditions of receiving waters and devise effluent standards that minimize costs, the regulatory task becomes insurmountable.[89] In a system of effluent charges or taxes, equity requires that dischargers pay for the use of the assimilative capacity of surface waters whether they are privately or collectively owned. Studies, it is claimed, show that industry can reduce waste discharges enormously with the proper incentive to do so.[90] Establishing a pollution charge or tax system in conjunction with water quality standards would resolve most of the

30

political conflict over the enforcement of water quality, and those affected could see exactly how they were affected.[91] Regulatory policy that imposes costs on industry and provides government subsidies to municipalities hides the costs of pollution control strategy.

Advocates of effluent charges or taxes also assert that these devices can be structured in different ways, but, properly structured, economic incentives would guide and decentralize decision making. Dischargers can minimize their costs by reducing the discharge of pollutants to the point where the marginal cost of reducing pollutants by another unit will just equal the effluent charge or tax. The lower the incremental cost of waste reduction, therefore, the higher the optimum removal percentage. A combined hydrological, engineering, and economic analysis of the Delaware River Basin concluded that discharge reduction to a given level of water quality could be achieved by use of effluent charges at a cost of 40 to 50 percent less than uniform treatment requirements by regulation.[92] Each waste discharger is charged in proportion to the use he makes of the assimilative capacity of surface waters, making the system more equitable. The charge scheme takes advantage of different control costs for different dischargers to keep total control costs down.[93]

Advocates of the effluent tax or charge claim that it would provide positive incentives to dischargers to achieve pollution reductions immediately, thereby reducing the burden on enforcement staff of government agencies.[94] Dischargers can select the most cost efficient way to achieve pollution reductions. Effluent charges or taxes are neutral as to the technique chosen. Advocates claim this sytem will reduce waste, not just its growth rate. A user charge by the city of Cincinnati is claimed to have reduced waste volume 40 percent in one year. Since firms that are least efficient in reducing wastes would sustain the highest costs, dischargers will have an incentive to reduce waste generation and to recycle wastes. This is a built-in incentive for firms to engage in research and development on waste reduction in order to survive in the marketplace.[95]

The effluent charge or tax approach to controlling water pollution would not strain an already overextended tax system; it would yield revenues and could get the federal government out of the business of financing local waste treatment facilities from tax revenues.[96] Advocates cite examples of the use of taxes and charges for controlling pollution both domestically and abroad.[97]

Opponents of effluent charges or taxes to control pollution in surface waters insist that no president could survive a proposal to vest property rights in the assimilative capacity of surface waters in industries. Such a system, it is claimed, would be too complex to administer. We currently lack sufficient information on water quality levels in receiving waters to provide certainty to the fee system. Enforcement authorities would not know what level of cleanup could be achieved unless dischargers' reactions to the system could be predicted accurately in advance. It would be most wasteful if dischargers made significant investments in response to the fee initially set and the fee were then changed because it did not produce the responses expected.[98]

Opponents also note that such a fee system would be a license to pollute. The environment, in their view, is not for sale or lease. There is political capital to be gained by cracking down on polluters through regulatory and enforcement actions but none in allowing industry to pollute for a fee. Some businesses contend the fee system would be unnecessary. Regulations are sufficient to secure compliance, in their view, and fees or taxes would amount to overkill.[99] Some opponents contend that an effluent fee or charge or an equivalent tax would take money away from business that might otherwise be spent on the cleanup effort. There are significant economies of scale in pollution abatement. Small businesses that lack economies of scale and the capacity for research and development would be hardest hit and might fail.[100]

Advocates of an effluent charge or tax respond that the level of fee or tax can be approximated, and smaller firms could be charged at a lower rate if that is necessary.

Marketable Permits. Advocates of marketable or tradable permits contend that this system of pollution control makes it easier to accommodate economic growth. An outside firm can buy permits that will allow it to establish a factory in a particular watershed, and no advance approval is needed for that purchase. Enforcement is easier than for effluent charges, it is claimed, as the authorities do not have to measure the totality of pollution. They can spot check a particular discharger to be sure it is within its quota as set by its permits. This strategy is said to be more certain in its effect on control of pollution than effluent charges.

A regulatory control strategy, as under the 1972 statute, forces some dischargers to overdo pollution abatement. Trading in permits would enable these firms to offset some of their costs. The cost at which permits are traded would influence pollution control expenditures. A firm wishing to minimize overall expenditures would elect to reduce pollution to the point where the marginal cost of abatement would equal the market price of a permit. Supporters consider that tradable permits minimize control costs and provide greater flexibility to businesses.[101]

Opponents of tradable permits contend this system still constitutes a license to pollute. There would be uncertainty, it is claimed, as to the price for a permit, and this would mean the process would not carry appropriate signals to the market about the social costs of pollution. Opponents fear that trading in permits would encourage an undue concentration of dischargers in a small locality, compounding pollution problems rather than solving them. Administration of the system of tradable permits is viewed as complicated. Experience in some areas indicates that there are too few proposals for trading permits to make a satisfactory market.

Subsidies. Advocates of using subsidies to encourage businesses not to pollute contend that taxes or effluent charges raise the prices of goods sold in the market as firms seek to recoup the amount of the charge or tax. Higher prices reduce the quantity of goods sold as well, leaving society as a whole worse off. Subsidies can accomplish the same results in pollution abatement as effluent charges or taxes, but without some of their disadvantages.[102]

Opponents contend that subsidies amount to a bribe not to pollute, and the government should not bribe businesses. Because such a bounty strategy depends for its success on knowledge of total emissions, it would be more difficult to administer than a marketable permit system. The sternest criticism is leveled at the incentives inherent in the system. Businesses would have little incentive to find innovative ways to reduce the discharge of pollutants. On the contrary there would be an incentive to generate pollutants until the cost of their treatment was less than the per unit subsidy. Firms might actually go into waste making to earn the subsidy. The burden of paying for this strategy would be imposed on taxpayers generally rather than on the polluting firms and their customers, thus raising concerns about equity.[103]

10

Deadlines And Exemptions

The administration's proposal would not change the basic strategy of the 1972 statute or its goals. It would extend deadlines for compliance with regulatory requirements of the statute. A discharge permit would be effective for a maximum of ten years instead of five.

Extension Of Deadlines

Advocates of liberal extensions for deadlines cite the statement of Douglas Costle who, as administrator of EPA, acknowledged that the task of setting standards was much more resource intensive and time consuming than had been expected.[104] Although EPA has issued guidelines for metal finishers and is under court order to publish BAT guidelines for all industries, its delay in issuing guidelines makes it impossible for industries to plan, build, and place in operation the facilities required to meet these guidelines within the time allowed by the statute. According to Senator Randolph, 1982 testimony showed that at least three years is needed after issuance of guidelines and the issuance of permits for industries to design, construct, and place in operation the facilities needed to comply with guideline requirements.[105] Advocates of extensions note that Congress has allowed municipalities until July 1, 1988, to comply with secondary treatment requirements for municipal wastewater plants. They argue that industry should be allowed at least as much time. Senator Randolph's bill, S. 432, calls for an extension of forty-two months from the date of enactment of amendatory legislation or thirty-six months after promulgation of guidelines or the issuance of permits, whichever period is longer.

Representatives of the steel and chemical industries contend that, if they are compelled to spend large sums of money to meet BAT requirements, they will be unable to develop new energy supplies through techniques such as turning coal into liquid fuel.[106] Gordon Wood, vice-president of the Olin Corporation, contends that BAT requirements are often superfluous—treatment for treatment's sake. "They require treatment just because the equipment is available, not because it is needed to make a river meet the public health standards."[107] Industry groups would thus go further than the administration and authorize EPA to waive BAT requirements if an industry can prove that continuing to treat discharges conventionally will not degrade the receiving waters.

J. Taylor Banks, attorney for NRDC, the organization that brought the successful suit to compel EPA to issue guidelines, contends that extension of additional time to comply would allow millions of pounds of toxic chemicals to be released each year.[108] NRDC favors a sliding scale requiring each industry to comply within thirty-six months after EPA issues its rule. Clint Whitney, executive director of the California Water Resources Control Board, argues that an extension to 1987 is adequate and that deadlines lose their value as action-forcing devices if they are too liberal.[109]

EXEMPTIONS

The adminstration's bill would modify the definition of pollutant to exclude munitions used in the course of conventional military weapons training and testing by the armed forces or by our allies in joint training exercises. Advocates of this change contend that we can ill afford to let the regulatory requirements of the statute interfere with the operational readiness of the military forces. The amendment is intended to permit the military to conduct training exercises without having to obtain a permit for the intentional or accidental discharge of projectiles into waters near or surrounding targets.[110]

The administration's proposed amendment was stimulated by a court decision holding that a permit to discharge munitions was required by the clean water legislation. While the Supreme Court held in *Weinberger* v. *Romero-Barcelo*, a case that arose in Puerto Rico, that a court injunction is not automatically required in the event of failure to obtain a permit, advocates of the amendment insist that our military readiness could be compromised by such requirements. Even an accidental discharge of munitions into or near water might violate the current law.[111]

Critics of the proposed change note that Congress explicitly made all three branches of the government subject to the requirements of clean water legislation. The government is frequently the greatest polluter of our surface waters. Munitions discharge, it is asserted, can be quite as damaging to the quality of water as other discharges. Military authorities know well in advance when and where they will conduct training exercises, and they can seek permits to accommodate such training exercises.

Advocates of the munitions exemption insist that our national security is too important to tie up military authorities in the permit process for months while awaiting the award of a permit and, indeed, the permit process was never intended for the discharge of munitions by our armed forces.

PERIOD OF PERMIT LIFE

The proposed legislation would extend the period of permit effectiveness from no more than five years to no more than ten years. The administration argues that this change will provide the certainty needed to encourage investments in pollution controls, will help eliminate a paperwork backlog, and will allow a more efficient use of resources.

Critics of this proposed change contend that the terms and conditions of permits should be reexamined frequently to ensure use of the appropriate technology and to determine whether other changes are warranted.

S. 431 would require EPA to impose more stringent control standards before the ten-year period is up if advances in technology produce more efficient pollution control mechanisms or if a new health hazard is discovered in chemicals contained in plant discharges. Ruckelshaus contends that EPA and states could, through regulations under the administration's bill, establish shorter maximum terms for categories of permits that should be reviewed more frequently because of such special concerns as evolving control requirements, compliance difficulties, and water quality effects.[112]

11

ENFORCEMENT

The administration's bill and S. 431 would permit EPA to impose administratively a series of penalties not exceeding $75,000 in the aggregate. Separate criminal sanctions would be provided for knowing and negligent violations of provisions of the statute, violations of permit conditions or limitations, and violations of requirements imposed in a pretreatment program or in a permit for deposit of dredged and fill materials, and for introducing pollutants or hazardous substances into a sewer system or municipal treatment plant. Knowing submission of a false material statement and knowing tampering with a monitoring device would also be made felonies.

Advocates of these changes argue that the administrative imposition of civil penalties would expedite enforcement efforts. If the violator wishes an administrative hearing, a neutral hearing officer would preside. Court review could be sought after the administrative hearing. Advocates note that the Administrative Conference of the United States several years ago recommended a similar approach to the adjudication of civil penalties to reduce the backlog of cases in federal courts and speed enforcement action. The increased penalties provided are needed, it is argued, as a deterrent to prohibited activities.

Opponents of these changes contend that, as the Supreme Court has noted, the clean water legislation already contains unusually elaborate enforcement provisions. Opponents fear that agency imposition of penalties will bring "kangaroo court" impositions. Hearing officers, judging by past experience, have often graduated from the ranks of the agency's employees and are hardly neutral.

Increased criminal sanctions are criticized as imposing harsh new sanctions for offenses already having adequate city, state, and national remedies. Dropping a thimbleful of pollutants into a commode and flushing it into the sewer system could be a violation of the proposed new provisions. Statutes on false statements are already found in title 18 of the United States Code, and no duplicating provision is needed. In sum, opponents view these additions as a clear case of overkill.

The Supreme Court has ruled that private litigants cannot bring actions against polluters based on the federal common law of nuisance since the statute preempts causes of action other than those set out in the act.[113] S. 431 provides, however, that litigants may sue under state common law, under other federal statutes, or under the clean water statute. Environmentalists contend that this approach will provide additional deterrence and honor our tradition of federalism.

Opponents contend that only one cause of action is needed. The statute adequately addresses remedial litigation concerns.

Scholars who advocate the use of economic incentives contend that providing such incentives could improve compliance with clean water objectives.[114] Such provisions might make increased criminal sanctions and added civil causes of action unnecessary.

12

ADMINISTRATION

The administration's bill would allow EPA to impose fees for processing certain applications. It would also make businesses constructing new plants subject only to new source performance standards actually promulgated by regulation in final form before construction is started. Currently businesses are subject to new source performance standards that are merely proposed before construction begins. Under section 1 of the bill EPA can relinquish a part of its permit-granting activities to states unable to assume full responsibility for all such activities.

Advocates of these changes claim that polluters should be required to pay for the cost of processing their applications. This is consistent with numerous proposals to require those dealing with the government to pay users' fees. The money, it is pointed out, would be earmarked for the particular function involved. The effective date of new performance standards would be adjusted as a matter of equity. There is nothing to prevent EPA from changing requirements between the date of proposal in regulation form and final issuance. Businesses therefore should not have to comply with requirements that may change and impose much unnecessary expense. Allowing states to proceed with even a part of permit-granting activities, it is claimed, would help relieve the severe strain on EPA employees and speed the process to the advantage of all.

Some businesses fear that EPA may set excessive fees for processing applications because the statute gives no guidance as to the amount of these fees. Small businesses may be particularly hard hit by fees. Some critics insist that businesses did not seek the permit process. That process is a governmental function, and tax revenues should fund its performance. Some environmentalists insist that new plants should be subject to new source performance standards from the date they are proposed in regulations. They contend that few changes are made before these regulations become final, and expeditious cleanup of our surface waters requires that plants not be allowed to avoid the latest technology by starting construction before a new source performance standard regulation becomes final. Some environmentalists fear that haphazard delegation of permit-granting authority to the states will mean that not all reviews of applications will be as thorough as they should be. States, in their view, should have a certifiable program for the entire activity before EPA relinquishes authority to them.

NOTES TO TEXT

1. Primary treatment is the first stage of wastewater treatment where substantially all floating and "settleable" solids are removed by flotation or sedimentation. 40 C.F.R. 125.58(m). Based on a monthly average, secondary treatment requires that the treated effluent shall have no more than forty parts per million of biochemical oxygen demand (BOD) and thirty parts per million of suspended solids, and that 85 percent of the BOD and suspended solids entering the plant be removed (also based on a monthly average). See House Committee Print 97-30, p. 2, note 1, and p. 9, note 5, for this standard and the sophisticated processes employed to achieve secondary standards.

2. U.S. Congress, Senate, Committee on Environment and Public Works, Subcommittee on Environmental Pollution, *Hearings on Clean Water Act* (hereafter *1982 Senate Hearings*), February 5, 1982, statement of John W. Hernandez, Jr., deputy administrator of EPA, p. 9. An EPA study concluded that secondary treatment at forty municipal treatment plants removed from 30 to 92 percent of the pollutants tested. U.S. General Accounting Office (GAO), *A New Approach Is Needed for the Federal Industrial Wastewater Pretreatment Program*, 1982, p. 6. See also, "Environment: Effluent Guidelines Rulemaking Ends," *Bureau of National Affairs Daily Report for Executives* (hereafter *BNA DER*), January 25, 1983, pp. C-1, C-3.

3. "Congress to Review Clean Water Legislation," *Congressional Quarterly*, vol. 40 (January 23, 1982), p. 124.

4. "Environment: Phaseout of Sewer Grants Program over 1-Year Period Urged," *BNA DER*, October 7, 1982, pp. A-3, A-4.

5. GAO, *A New Approach Is Needed*, pp. 5–6.

6. Ibid., pp. 3–4.

7. Ibid., p. 1.

8. Cited in Morris A. Ward, *The Clean Water Act: The Second Decade* (Washington, D.C.: E. Bruce Harrison Co., 1982), p. 33.

9. GAO, *A New Approach Is Needed*, pp. 10–11; "Congress to Review Clean Water Legislation," p. 124.

10. William Chapman, "Environmental Groups Assail EPA on Its Clean-Water Proposals," *Washington Post*, April 9, 1982; "Congress to Review Clean Water Legislation," p. 124.

11. Frances Dubrowski, "EPA's Clean Water Bill: A Formula for Environmental Failure," *Federal Bar News and Journal*, vol. 29 (September/October 1982), pp. 327, 328.

12. Cited in Ward, *The Clean Water Act*, p. 34.

13. "Environment: No Mid-Course Correction Needed in Clean Water Act, Chafee Tells EPA," *BNA DER*, July 23, 1982, pp. A-2, A-4.

14. GAO, *Billions Could Be Saved through Waivers for Coastal Wastewater Treatment Plants*, 1981, p. 2.

15. Senator Christopher Dodd, "Ocean Waivers," *Congressional Record*, vol. 129 (June 23, 1983), p. S 9006.

16. GAO, *Billions Could Be Saved,* pp. 7–8.

17. U.S. Environmental Protection Agency, *Issues Concerning the Implementation of the Clean Water Act*, 1982, pp. 44–45.

18. Senator John G. Tower, "Legislation Relating to Discharge of Dredged and Fill Material," *Congressional Record*, vol. 127 (March 24, 1981), p. S 2538.

19. Ibid., p. S 2585.

20. *1982 Senate Hearings*, February 5, 1982, statement of Hernandez, p. 15; U.S. General Accounting Office, *Better Monitoring Techniques Are Needed to Assess the Quality of Rivers and Streams*, 1981, p. iii.

21. David P. Currie, "Congress, the Court, and Water Pollution," *Supreme Court Review*, vol. 1977, pp. 39, 61.

22. U.S. General Accounting Office, *Environmental, Economic, and Political Issues Impede Potomac River Cleanup Efforts*, 1982, pp. 25, 75, 81.

23. House Committee Print 97-30, October 1981, p. 73. A criterion issued by EPA is based solely on data and scientific judgment and does not reflect considerations of technical feasibility. U.S. Council on Environmental Quality, *Environmental Quality—1979*, p. 144.

24. Allen V. Kneese and Charles L. Schultze, *Pollution, Prices, and Public Policy* (Washington, D.C.: Brookings Institution, 1975), pp. 82–83.

25. "Moves That May Dilute the Clean Water Act," *Business Week*, June 21, 1982.

26. The court in API v. EPA, 661 F. 2d 340 (5th Cir. 1981) referred to this policy as analogous to a strict liability standard in tort law; that is, a discharger is responsible for incurring major costs of wastewater treatment even though its discharges cause no harm to the environment. EPA personnel considered amendments to the statute that would allow EPA to issue seasonal permits tailoring pollution control requirements to the assimilative capacity of the receiving waters at different times of the year. "Environment: EPA Draft Water Act Reforms Include Seasonal, Variable NPDES Permits," *BNA DER*, January 21, 1983, pp. A-1, A-2.

27. Richard B. Stewart, "Pyramids of Sacrifice?" *Yale Law Journal*, vol. 86 (May 1977), pp. 1196, 1220.

28. A. Myrick Freeman III, "Air and Water Pollution Policy," in Paul R. Portney, ed., *Current Issues in U.S. Environmental Policy* (Washington, D.C.: Resources for the Future, 1978), pp. 49, 51, 55–56.

29. Some critics question the need for both phosphorus and nitrogen controls, for example, because in theory limiting either one of these could control algae. GAO, *Environmental, Economic, and Political Issues Impede Potomac River Cleanup Efforts*, p. iii; Freeman, "Air and Water Pollution Policy," p. 73.

30. See Senate Report 92-313 accompanying the 1972 legislation that installed the national technology-based standards approach. See also, "Moves That May Dilute the Clean Water Act," quoting Thomas C. Jorling, former EPA administrator for water and hazardous wastes.

31. Currie, "Congress, the Court, and Water Pollution," pp. 60–61.

32. Limited provision was made for waiver of secondary treatment in the case of certain ocean outfalls. Statutory and EPA restrictions on greater use of this waiver authority have been heavily criticized.

33. "Environment: 1983 Water Quality Goal Cannot be Met," *BNA DER*, July 17, 1979, p. A-15, citing testimony of Wilbur D. Campbell, assistant director of GAO's Community and Economic Development Division; House Committee Print 97-30, p. 34.

34. A. Myrick Freeman III and Robert J. Haveman, "Clean Rhetoric and Dirty Water," *Public Interest*, no. 28 (Summer 1972), pp. 51–55; compare David M. Shell, "Skagway v. EPA: Cassandra's Prophecy Revisited," *Regulation: AEI Journal on Government and Society*, vol. 4 (Nov./Dec. 1980), p. 50, as to the lack of need for an expensive new treatment facility.

35. Kneese and Schultze, *Pollution, Prices, and Public Policy*, pp. 36–37.

36. Freeman, "Air and Water Pollution Policy," p. 61.

37. Ibid., p. 55; Kneese and Schultze, *Pollution, Prices, and Public Policy*, p. 43.

38. U.S. Congressional Research Service, *Municipal Pollution Control: The EPA Construction Grants Program*, 1982, p. 5; and see Patrick E. Taylor, "Dirty Water: Federal Failure Part II," *Washington Post*, May 11, 1981.

39. U.S. General Accounting Office, *Costly Wastewater Treatment Plants Fail to Perform as Expected*, 1980, p. 9.

40. Kneese and Schultze, *Pollution, Prices, and Public Policy*, p. 37.

41. Freeman and Haveman, "Clean Rhetoric and Dirty Water," p. 63.

42. See Senate Report 97-204.

43. U.S. Council on Environmental Quality, *Environmental Quality—1980*, p. 100.

44. U.S. General Accounting Office, *National Water Quality Goals Cannot be Attained without More Attention to Pollution from Diffused or "Nonpoint" Sources*, 1977. Counsel directly involved in clean water issues at EPA when the 1972 amendments were adopted attributes failure of the legislation to provide for adequate control of nonpoint source pollution to the fact no one had a good idea of how federal control of such pollution could be achieved. Robert Zenor, "The Federal Law of Water Pollution Control," in Erica L. Dolgin and Thomas P. Guilbert, eds., *Federal Environmental Law* (St. Paul, Minn.: West Publishing Co., 1974), pp. 682, 769.

45. See Freeman, "Air and Water Pollution Policy," pp. 12, 52. EPA Administrator Ruckelshaus cites the 1982 state water quality assessments report that in one-fifth of the states nonpoint sources of pollution are now the most important cause of water degradation in waters not meeting designated uses. Agricultural nonpoint sources are the major sources of degradation in lakes with eutrophication problems. Twenty states specifically quantified their progress toward the statute's goals. Fifteen of these states list nonpoint sources of pollution as significant sources of pollution in their remaining problem waters. U.S. Congress, Senate, Committee on Environment and Public Works, Subcommittee on Environmental Pollution, *Hearings on Reauthorization of the Clean Water Act*, June 14, 1983, statement of Ruckelshaus, p. 13.

46. GAO, *A New Approach Is Needed*, p. 6. A Department of Commerce study cited in the GAO report states that in the case of copper, zinc, and cadmium, residential sources contribute more than electroplaters. Except in the case of nickel, storm-water runoff contributes more to the city picture than electroplaters.

47. "Environment: 1983 Water Quality Goal Cannot be Met," p. A-15. GAO estimated that as many as thirty-seven states might not attain water quality standards and goals without greater attention to pollution from nonpoint sources. The 1976 report of the National Commission on Water Quality recommended extending controls to urban storm runoff and agricultural and other nonpoint sources.

48. EPA estimates that 795 of the 3,977 major nonmunicipal dischargers did not meet the best practicable treatment deadline.

49. CEQ, *Environmental Quality—1979*, p. 76.

50. *1982 Senate Hearings*, February 5, 1982, statement of Hernandez, p. 7.

51. Senate Report 97-204, p.2.

52. Joseph J. Seneca and Michael K. Taussig, *Environmental Economics* (Englewood Cliffs, N.J.: Prentice-Hall, Inc., 1974), p. 122. Environmentalists acknowledge that EPA does not consider the quality of receiving waters in setting technology-based standards. States, they point out, may consider the quality of receiving waters in setting water quality effluent limitations more strict than the standards set by EPA. This permits states to protect designated uses of particular bodies of water when they feel EPA standards do not go far enough to eliminate the discharge of pollutants. State-designated uses typically include: Class A—drinking use of water; Class B—fishing and swimming; Class C—fishing and boating but not water contact sports; and Class D—not for fishing or boating.

53. See Senate Report 92-414, committee findings; Senate Report 96-744, p. 2.

54. Senate Report 92-414, discussion of section 207 grant authorization.

55. These data do not include pre-1972 expenditures. Federal appropriations for the construction of public wastewater treatment plants alone during the period from 1957 through 1971 totaled over $3 billion. U.S. Environmental Protection Agency, *Clean Water—Report to Congress*, 1973, p. 38.

56. CEQ, *Environmental Quality—1980*, table 10-2, p. 397. Public law 97-117 cut the percentage of the federal grant contribution to future public treatment plants. This may slow the rate of future expenditures because of the reluctance of local authorities to shoulder expensive new projects when they must pay a higher percentage of total costs. In the 1976 report of the National Water Commission it estimated that best technology available would require the expenditure of $470 billion over a ten-year period of time. Allen V. Kneese, "Costs of Water Quality Improvement," in Henry M. Peskin and Eugene P. Seskin, eds., *Cost-Benefit Analysis and Water Pollution Policy* (Washington, D.C.: Urban Institute, 1975), pp. 175, 181. For a $228 billion estimate (in constant 1982 dollars) for spending from 1980–1989 to reach standards set in the statute for both industry and government, see "Cleaner Air and Water," *U.S. News & World Report*, vol. 94 (February 28, 1983), p. 27.

57. CRS, *Municipal Pollution Control*, p. 4. Transition team papers submitted to the incoming Carter administration in 1976 stated that elimination of all municipal pollutant discharges as required by the 1972 statute could cost $444 billion. "Sewers, Clean Water, and Planned Growth," *Yale Law Journal*, vol. 86 (March 1977), pp. 733, 735. See also "Sewage Plant Needs," *Washington Post*, January 4, 1983.

58. GAO, *Environmental, Economic, and Political Issues Impede Potomac River Cleanup*, pp. ii, 65–72, 77, 94–99; Daniel Rapoport, "Sludge," *National Journal*, vol. 8 (September 11, 1976), p. 1284; see also, Mancur Olson and Hans H. Landsberg, *The No-Growth Society* (New York: W.W. Norton Co., 1973), as to the problem of side effects. A Department of Commerce study suggests that environmental controls could consume as much as 8.2 percent of energy industry output by 1983. An EPA study suggests that pollution abatement accounted for approximately 2 percent of national energy use in 1977, and this figure may grow to 3 percent within a decade. U.S. Congressional Research Service, *Environmental Regulations: Economic Impact*, 1982, p. 12.

59. "Clean-Water Campaign Springs Some Leaks," *U.S. News & World Report*, vol. 86 (December 24, 1979), pp. 59, 60.

60. CEQ, *Environmental Quality—1978*, pp. 144—146; EPA, *Issues Concerning Implementation of the Clean Water Act*, p. 5. For the special problems of an Alaskan village of 850, see Shell, "Skagway v. EPA," p. 50.

61. Barnaby J. Feder, "Technology Standards for Pollutants," *New York Times*, April 15, 1982.

62. See the critique of methodologies employed in various benefit studies in Dennis Tihansky, "A

Survey of Empirical Benefit Studies," in Peskin and Seskin, eds., *Cost-Benefit Analysis and Water Pollution Policy*, p. 127.

63. A. Myrick Freeman III, *The Benefits of Air and Water Pollution Control: A Review and Synthesis of Recent Estimates*, a report prepared for CEQ, December 1979. A typical methodology is to estimate the value per day to users of the activity for which a recreational site will be used and then estimate in days the increased demand that cleaner water will produce. Ibid., pp. 72, 85–95.

64. Freeman, "A Survey of the Techniques for Measuring the Benefits of Water Quality Improvement," p. 72; John Bishop and Charles Cicchetti, "Some Institutional and Conceptual Thoughts on Measurement of Indirect and Intangible Benefits and Costs," in Peskin and Seskin, *Cost-Benefit Analysis and Water Pollution Policy*, pp. 105, 134, 150—51.

65. See U.S. Congressional Research Service, *Impact of Current Budget Cuts on EPA's Construction Grants Program*, 1982, p. 5. The National Utility Contractors Association places a much higher estimate on the employment value and other stimulative effects of the government's expenditures for new sewer plants. See Lawrence Mosher, "Clean Water Requirements Will Remain Even if the Federal Spigot Is Closed," *National Journal*, vol. 13 (May 16, 1981), p. 874.

66. Robert H. Haveman and Burton A. Weisbrod, "The Concept of Benefits in Cost-Benefit Analysis: With Emphasis on Water Pollution Control Activities," in Peskin and Seskin, eds., *Cost-Benefit Analysis and Water Pollution Policy*, p. 44; CRS, *Environmental Regulations: Economic Impact*, p. 10.

67. See Allen V. Kneese and Blair T. Bower, *Managing Water Quality: Economics, Technology, Institutions* (Washington, D.C.: Resources for the Future, 1968), p. 178; Feder, "Technology Standards for Pollutants" (electroplaters claim best practicable technology would put half of them out of work). Paint industry officials, per GAO, claim they will have to close their plants if the zero discharge standard is not changed (GAO, *A New Approach Is Needed*, p. 7). A study by EPA found that between January 1971 and January 1982 pollution control costs were claimed to be a factor in fifty-four plant closings. An additional fifteen plant closings involved complaints with both clean air and clean water requirements. "Environment: Pollution Control Costs Said to Close 154 Plants since 1971," *BNA DER*, January 6, 1983, p. A-7. Overall the study estimated that air and water pollution control costs may have been a factor in the closing of 154 plants employing 32,749 persons.

68. Aaron Wildavsky, " Economy and Environment: Rationality and Ritual," *Stanford Law Review*, vol. 29 (November 1976), pp. 183, 187.

69. Compare Freeman, *The Benefits of Air and Water Pollution Control*, pp. 163–65, with Freeman, "A Survey of the Techniques for Measuring the Benefits of Water Quality Improvements," p. 95.

70. Haveman and Weisbrod, "The Concept of Benefits in Cost-Benefit Analysis," p. 38. Economists and business persons caution that special care must be taken in evaluating benefits of cleaner water because action that may improve water for one use may degrade it for another. Thermal pollution may hurt sport fishing for trout or salmon but improve water for swimming. Industrial acids may affect fishing but retard the growth of algae. People may tolerate a pH as low as 5.5 or 5.0 in any event. Tihansky, "A Survey of Empirical Benefit Studies," pp. 147, 153, 159. Acid damage, it is claimed, is generally confined to a few sites where water use is extensive. Less oxygen in water may not be important to many people as they may only be peripherally aware of the deficiency. Though oxygen is important to other people, less oxygen may be beneficial to industry for oxygen corrodes intake facilities at industrial plants. Water hardness and dissolved solids may be due to natural causes and not to point source dischargers. It is thus easy for those putting values on the clean water initiative to overestimate benefits in this area. In addition, it is asserted, use values are assigned to cleaner surface waters when recreational users might find it more satisfying and less expensive to use alternative sites nearby.

71. CRS, *Environmental Regulations: Economic Impact*, p. 11, citing CEQ, Brookings Institution, and Data Resources, Inc., that all environmental expenditures increase the inflation rate by 0.3 to 0.5 percent a year. For the claim that pollution abatement expenditures contribute to heavy demands on the money market which keeps interest rates high, see U.S. General Accounting Office, *Environmental Protection: Agenda for the 1980s*, 1982, p. 31.

72. Compare CEQ, *Environmental Quality—1980*, pp. 387–89.

73. "Laws Not Blocking Projects, Not Causing Location Abroad, Reports Says," *BNA, DER*, November 15, 1982, p. A-1.

74. GAO, *Environmental Protection: Agenda for the 1980s*, p. 31.

75. Freeman, "Air and Water Pollution Policy," p. 19; GAO, *Environmental Protection Issues in the 1980s*, 1980, p. 2; GAO, *Environmental, Economic, and Political Issues Impede Potomac River Cleanup*, p. 107.

76. Haveman and Weisbrod, "The Concept of Benefits in Cost-Benefit Analysis," pp. 39–40.

77. Lester Lave, "Environmental Solutions Aren't Simple," *Washington Post*, April 6, 1983.

78. Kneese and Schultze, *Pollution, Prices, and Public Policy*, pp. 19–23; Freeman and Haveman, "Clean Rhetoric and Dirty Water," p. 59; "Controversy over Costs," *National Journal*, vol. 4 (January 22, 1972), p. 173.

79. Freeman, "Air and Water Pollution Policy," p. 49.

80. Compare the analysis made at an earlier date on the same estuary in James C. Hite, Hugh H. Macaulay, James M. Stepp, and Bruce Yandle, Jr., *The Economics of Environmental Quality* (Washington, D.C.: American Enterprise Institute, 1972), pp. 19–23.

81. See Stewart, "Pyramids of Sacrifice?" p. 1199; "Water Quality Report," *1976 Congressional Quarterly Almanac* (Washington, D.C.: Congressional Quarterly Publishing Co., 1976), pp. 170–71.

82. GAO, *Environmental, Economic, and Political Issues Impede Potomac River Cleanup*, p. ii.

83. Mark Sagoff, "Economic Theory and Environmental Law," *Michigan Law Review*, vol. 79 (June 1981), pp. 1393, 1394, 1408, 1410.

84. Senate Report 92-414.

85. See Zenor, "The Federal Law of Water Pollution Control," p. 698.

86. "Environment: Bubble Policy May Be Extended Next to Water Pollution," *BNA DER*, May 6, 1982, pp. A-6, A-7, citing statement of Frances Dubrowski of NRDC. Dubrowski claims that six analyses prepared for EPA rules under the executive order requiring such analyses averaged $539,000 per rule. "Environment: EPA, Environmental Groups Oppose Hearings in Regulatory Reform Proposed Legislation," *BNA DER*, June 29, 1983, pp. A-24, A-25.

87. See, for example, "Environment: Superfund Controversy, Other Issues Stall Action on Clean Water Act Changes," *BNA DER*, March 11, 1983, pp. A-5, A-6, quoting Willen Van der Broek, staff aide to the Water Resources Subcommittee of the House Committee on Public Works and Transportation.

88. "Environment: Groups Spar over Clean Water Act Proposal to Allow Local Pretreatment Programs," p. A-8, and "Environment: House Panel Urged to Maintain Funding," *BNA DER*, August 9, 1982, pp. A-4, A-5.

89. Kneese and Schultze, *Pollution, Prices, and Public Policy*, p. 91.

90. Allen V. Kneese, "Discharge Capacity of Waterways and Effluent Charges," in Selma J.

Muskin, ed., *Public Prices for Public Products* (Washington, D.C.: Urban Institute, 1972), pp. 133, 147.

91. Freeman and Haveman, "Clean Rhetoric and Dirty Water," p. 53.

92. Kneese and Schultze, *Pollution, Prices, and Public Policy*, p. 81; Kneese, "Costs of Water Quality Improvement, Transfer Functions, and Public Policy," p. 183. See CRS, *Environmental Regulations: Economic Impact*, p. 12, for the view of the Council on Wage and Price Stability that substitution of a fee system would significantly reduce costs. See also, GAO, *Environmental Protection: Agenda for the 1980s*, p. 29.

93. Frederick R. Anderson, Allen V. Kneese, Phillip D. Reed, Russell B. Stevenson, and Serge Taylor, *Environmental Improvement through Economic Incentives* (Washington, D.C.: Resources for the Future, 1977), p. 33; Allen V. Kneese and Blair T. Bower, *Managing Water Quality: Economics, Technology, Institutions* (Washington, D.C.: Resources for the Future, 1968), p. 141.

94. Freeman, "Air and Water Pollution Policy," p. 51; Anderson et al., *Environmental Improvement through Economic Incentives*, p. 34. Laurence H. Tribe, "Too Much Law, Too Little Justice," *Atlantic*, vol. 244 (July 1979), pp. 25, 27. For use of the bubble approach for firms with multiple effluent outfalls, some exceeding effluent limits but with offsetting savings at other outfalls, see "Environment: Steelmakers, NRDC, EPA Agree on 'Bubble' for Waste Water Discharges," *BNA DER*, March 7, 1983, p. A-4.

95. Olson and Landsberg, *The No-Growth Society*, p. 232; Seneca and Taussig, *Environmental Economics*, p. 229; Kneese and Schultze, *Pollution, Prices, and Public Policy*, p. 92; Freeman and Haveman, "Clean Rhetoric and Dirty Water," p. 56.

96. Kneese and Schultze, *Pollution, Prices, and Public Policy*, p. 93; Kneese and Bowen, *Managing Water Quality*, p. 174.

97. Anderson et al., *Environmental Improvement through Economic Incentives*, pp. 59–66; Seneca and Taussig, *Environmental Economics*, pp. 237–39; Kneese and Bower, *Managing Water Quality*, pp. 237–45; Kneese and Schultze, *Pollution, Prices, and Public Policy*, pp. 109–11.

98. GAO, *Environmental Protection Issues in the 1980s*, p. 11; Freeman and Haveman, "Clean Rhetoric and Dirty Water," p. 54.

99. Seneca and Taussig, *Environmental Economics*, pp. 231, 233.

100. Anderson et al., *Environmental Improvement through Economic Incentives*, pp. 160–61.

101. Compare Clifford S. Russell, "Controlled Trading of Pollution Permits," *Environmental Science and Technology*, vol. 15 (January 1981), p. 24; and see the strategies considered by EPA staff including trading, banking, and trading between point and nonpoint sources. "Environment: EPA Draft Water Act Reforms Include Seasonal, Variable NPDES Permits," p. A-2.

102. Seneca and Taussig, *Environmental Economics*, pp. 222–23.

103. Ibid., Kneese and Bower, *Managing Water Quality*, pp. 177, 316.

104. Ward, *The Clean Water Act*, p. 31.

105. Senator Jennings Randolph, "Clean Water Compliance Date Extension Act of 1983," *Congressional Record*, vol. 129 (February 3, 1983), p. S 1011.

106. Kathy Koch, "Administration Proposals to Revise Clean Water Act Delayed by Objections," *Congressional Quarterly*, vol. 40 (April 17, 1982), p. 874; Kathy Koch, "Congress to Review Clean Water Legislation," *Congressional Quarterly*, vol. 40 (January 23, 1982), p. 124; William Chapman, "Environmental Groups Assail EPA on Its Clean-Water Proposals," *Washington Post*, April 9, 1982; Joanne Omang, "Industry Told of Easier Clean-Water Rules," *Washington Post*, January 13, 1982.

107. Koch, "Administration Proposals to Revise Clean Water Act Delayed by Objections," p. 874; Koch, "Congress to Review Clean Water Legislation," p. 124. The National Commission on Water Quality recommended in its 1976 report that Congress should allow exemptions from technology requirements where the discharger can demonstrate that installation of the new technology would produce minimal benefits ("Water Quality Report," p. 170).

108. Chapman, "Environmental Groups Assail EPA on Its Clean-Water Proposals."

109. "Environment: House Panel Urged to Maintain Funding," p. A-4. See, "Environment: Effluent Guidelines Rulemaking Ends," p. C-1, for NRDC and sliding scale.

110. "Environment: Need for Broader Military Exemption to Clean Water Act Questioned in House Subcommittee," *BNA DER*, July 28, 1982, p. A-1.

111. See Fred Barbash, "High Court Says Federal Judges May Delay in Fouled-Water Cases," *Washington Post*, April 28, 1982; "Environment: Clean Water Act Doesn't Require Injunction of Navy's Training, Supreme Court Says," *BNA DER*, April 28, 1982, p. A-10.

112. *Senate Hearings on Reauthorization of the Clean Water Act*, June 14, 1983, statement of Ruckelshaus, p. 11.

113. Middlesex County Sewerage Authority v. Sea Clammers, 453 U.S. 1 (1981).

114. Lave, "Environmental Solutions Aren't Simple."